YOUR KNOWLEDGE HAS VALUE

Bibliographic information published by the German National Library:

The German National Library lists this publication in the National Bibliography; detailed bibliographic data are available on the Internet at http://dnb.dnb.de .

Imprint:

Copyright © 2017 GRIN Verlag, Open Publishing GmbH
Print and binding: Books on Demand GmbH, Norderstedt Germany
ISBN: 9783668448230

This book at GRIN:

http://www.grin.com/en/e-book/359506/laboratory-preparation-and-analyses-of-ochre-soaps-with-characteristic

Adekunle Jelili Olaoye, Dauda B. E. N., Adedokun A. K.

Laboratory Preparation and Analyses of Ochre Soaps with Characteristic Medicinal Effect on Dermatophylosis

GRIN Publishing

GRIN - Your knowledge has value

Since its foundation in 1998, GRIN has specialized in publishing academic texts by students, college teachers and other academics as e-book and printed book. The website www.grin.com is an ideal platform for presenting term papers, final papers, scientific essays, dissertations and specialist books.

Visit us on the internet:

http://www.grin.com/

http://www.facebook.com/grincom

http://www.twitter.com/grin_com

LABORATORY PREPARATION AND ANALYSES OF OCHRE SOAPS WITH CHARACTERISTIC MEDICINAL EFFECT ON DERMATOPHYLOSIS

B.E.N Dauda1, A.J Olaoye 1 and K. A Adedokun3

(1) Department of chemistry Federal university of technology Minna.

(2) King Saud University Medical City, Riyadh

Content

ABSTRACT

Four oil namely palm kernel oil, palm oil, ground nut oil and shear butter oil were employed in the production of ochre soap. Dermatophilus congolensis were used as organism to test for anti-bacteria or medicated activities of this soap. The physical properties of the soap samples such as colour, texture and foaming were determine. The chemical properties such as acid value, percentage free alkalinity, percentage insoluble matter, percentage moisture and pH value were determined. It was found that the acid values of the soap with respect to oil were; 2.04, 2.12, 2.88 and 2.76 for palm kernel oil, palm oil, ground nut oil and shear butter oil respectively. The percentage free alkalinities of the soap with respect to oil were; 7.59, 7.69, 8.62 and 8.04 for palm kernel oil, palm oil, ground nut oil and shear butter oil respectively. The percentage moisture of the soap with respect to oil was; 2.4, 2.4, 5.4 and 2.8 for palm kernel oil, palm oil, ground nut oil and shear butter oil respectively. The percentage insoluble matter of the soap was; 2.2, 5.8, 4.4 and 6.6 for palm kernel oil, palm oil, ground nut oil and shear butter oil respectively. The pH values of the soap with the respect to oil were 9.00, 9.30, 9.10 and 9.40 for palm kernel oil, palm oil, ground nut oil and shear butter oil respectively.

BACKGROUND: There is paucity of data on Ochre substance for the treatment of dermatophylosis. Little is known about association between dermatophylosis and mud therapy where this orthodox medicine is still in practice. Large affected people by this causative agent, *Dermatophilus congolensis,* are usually asymptomatic carriers. In the affected people, no specific treatments have been reported. This current study aimed at Ochre soap preparation and laboratory analysis as well as exploring its medicinal effects on the microbial agent.

This was done for palm kernel oil, palm oil, groundnut oil and Shea butter oil respectively. Acid value and free alkalinity were titrated, with the percentage moisture content and pH values were all calculated. A pure isolate of *Dermatophylus congolensis* was sub-cultured on nutrient agar with holes bored on the agar using sterile cork-borer of 4.0 mm in diameter and $0.2cm^3$ of each prepared soap in liquid form was allowed to diffuse in an anaerobic jar for 48 hours. After incubation, the diameters of zones of inhibition were measured to the nearest millimeter.

KEYWORDS: Antimicrobial effect, Dermatophilosis, Ocher soap, Orthodox medicine, Mud therapy.

INTRODUCTION

Many prehistoric traditional doctrines of orthodox medicine particularly geophagy (the practice of eating earthy substances such as clay and chalk habitually believed to augment a mineral-deficient diet) have evolved in lots of discoveries useful in modern-day medicine. One particular ancient but innovative ideology is the use of ochre (an earth pigment containing silica, aluminum and ferric oxide) topically applied on body surface which was formerly known as mud therapy (Isabelle and Fabien 2012).

Ochre substances have been applied in several dermatologic infections on humans and animals, most frequently cattle, horses, sheep, and goats. However, little is known about ochre substance in the treatment of dermatophylosis and even underemphasized. Worsened more, the largely affected people are usually asymptomatic carriers (Abu-Samra, 1978). *Dermatophilus congolesis* is facultative aerobic actinomycetes causing a disease known as dermatophilosis (Albrecht *et al.*, 1974). Very few cases of dermatophilosis have been reported in humans as a result of poor diagnostic tools. The transmission mechanism of dermatophilosis is not yet clearly understood, however, mechanical transfer by zoonosis is a strong possibility. Previous skin lesion also appears to bring about the infection.

More importantly, several reports have shown that the infection origin remains unknown (Burd, *et al.*, 2007,). Also, the pathologic basis of this infection has implications on involvement of the living epidermis in a number of cases such as exudative and scaling lesion, pustular, folliculitis, tongue hairy leukoplakia, recalcitrant verruca, pitted keratosis, chronic nodular disease as well as in asymptomatic carriers (Abrose, 1996).

Currently, in humans no specific treatment has been reported for dermatophilosis, only that domesticated animals have been treated using mainly topical and parenteral antibiotics including other local preparations and have all proved practically ineffective. Apart from animals, human infections that have been self-limiting are now reported recurring in a number of cases (Burd, *et al.*, 2007).

However, possibility of use of modern medicated soaps not targeted for dermatophilosis is now a question; more so, these chemical agents remain in host-tissue safety doubt. Soaps are the alkali metal salt of long chain carboxylic acid, such as hexadecanoic (palmitic acid) and octadecanoic (stearic acid) (Ainie *et al.*, 1996). The carboxylic acid contains carbon between 12 and 18 carbon atoms (Mak- Mensah and Firempong *al.*, 2011). The nature (length of the chain) of the fatty acid largely determines the unique properties of soap (Mak- Mensah and Firempong *al.*, 2011). Fewer carbon atoms would result in soaps with skin irritation, while longer carbon chains would consequently bring about insoluble soaps. From animal and vegetable oils from which soaps are produced consist essentially of glycerol ester (Oguntola, 2002).

In common parlance, soaps are of various kinds and termed medicated, detergent, bar or even toilet soaps. Medicated soaps contain some other antimicrobial substances of medicinal importance. These medicinal substances from which medicated soaps are made may be sourced from plants, animals or natural resources (Mak- Mensah and Firempong *al.*, 2011). Some of the untapped natural resources are ochre substances. These soaps are specifically made for treatment of skin diseases or infections by applying them against the affected parts (Edgar, 1995). Skin infections occur as a result of bacteria, fungi, viruses or parasites which penetrate the body epidermal layer and innate defense system (George, 1984). Often times, most of these agents are susceptible to chemical agents. Up till date, little is known about antimicrobial effect of ochre agent. One of the goals of the current study was to employ the use of ochre substance in the preparation of laboratory-made medicated soaps from different oils and eventual testing on a pure strain of *Dermatophylus congolensis* obtained from research laboratory.

Action of soaps

Soaps are the most common cleansing reagents in use because of the surface activities of their molecules. These molecules are arranged in a regular patterns or cluster called micelles, mainly at the surface but sometimes throughout the body of the solution (Kuntom et al., 1994). Thus, soaps

- reduce the surface tension between the water and the oil or other insoluble materials,

- foam easily and

- stabilize the insoluble particles which are removed from the skin or clothes.

Soap have molecular structure with the characteristic of all surface active agents often called surfactant, each molecule contain a hydrophobic or water repellant component and a hydrophilic or water attracting component (Garzena and Marina, 2013).

CH$_3$. (CH$_2$)$_{16}$_____: COO-

Hydrophobic   : Hydrophilic

A soap molecule has an amazing characteristic. One end of the molecule is attracted to fats and repels water and the other end of the soap molecule with two oxygen atoms and one sodium atom are detracted to water, but repel fat. Molecules of water are attracted by the hydrophilic component, while the molecules of insoluble materials are attracted by the hydrophobic components (Garzena and Marina, 2013).

Ochre: Untapped Chemotherapeutic Agent in Orthodox Medicine

Ochre has so many benefits to offer as a natural product. It is used as anti-wrinkle and as a skin smoother especially for babies, and particularly exfoliates dead skin cell which on exposing the fresh skin underneath can allow it to breathe. It has also been scientifically demonstrated that ochre has antifungal and anti-inflammatory properties that are beneficial to the skin (Isabelle and Fabien 2012). These are earthly pigments of various colours ranging from yellow to brown-red. The most yellow varieties are minerals limonite likened with clay and may contain 16-65% Iron III oxide [Fe$_2$O$_3$].Yellow ochre substances are mined in France and South Africa notably. The redder varieties are akin to hematite and are found in Nigeria **(references)**.

MATERIALS AND METHODS

MATERIALS:

A. Pure culture of *Dermatophylus congolensis,*

B. Soap samples,

C. Apparatus such as;

1. Sterile petri dishes

2. Bunsen burner

3. Autoclave

4. Hot air oven

5. Conical flask

6. Test tubes and test tube rack

7. Aluminum foil paper

8. Masking tape

9. Incubator and

10. Disinfectant (dettol), and

D. Microbial culture media (for sensitivity);

    1. Nutrient Agar (NA)

METHODS:

Procedure for production of ochre soap

Sixty cubic centimeters ($60cm^3$) of different oils involving palm kernel oil, palm oil, groundnut oil, and shea butter oil was measured into four different $200cm^3$ beakers. Each was heated to melting point and poured into separate plastic containers. Also, one hundred and fifty cubic centimeters ($150cm^3$) of fermented sodium hydroxide solution was measured with measuring cylinder and added into the plastic containers. The mixture was stirred continuously to give a homogenous form. Likewise, $50cm^3$ of sodium carbonate was added into the plastic containers and the mixture was stirred. In addition, $60cm^3$ of sodium tripolyphosphate was added into the solutions followed by the addition of $37.5cm^3$ of sodium silicate. The solutions were stirred thoroughly and 2g of Ochre was later added, stirred continuously to achieve homogeneity.

Finally, Banana perfume and other additives such as sodium sulphate, sodium carboxyl methyl cellulose, and borax were added to the mixture. About 5mg of sodium chloride was added to it to salt out the soap. The whole mixture was stirred thoroughly and transferred into a mould. The soap was left for four days to solidify before it was removed from the plastic mould. This was done for palm kernel oil, palm oil, groundnut oil and Shea butter oil respectively **(Dunn, 2010)**.

Post-analytical procedure on finished ochre soap

1. Determination of acid value:

Calculation;

Acid Value   (A.V) $= \dfrac{a \times N \times 40}{P}$

a= Volume of sodium hydroxide

N= Molarity of the solution

P= Quantity taken for analysis (AOC, 1980).

2. Determination of % free alkalinity

Calculation;

Free alkalinity: $\dfrac{A \times N \times E \times 100}{P \times 1000}$

A = Volume of acid used

N = Molarity of solution

E = Equivalent weight of sample

P= Quantity taken for analysis (AOC, 1980).

3. Determination of % moisture content:

Calculation;

$$\% \text{ moisture in soap} = \left(\dfrac{Initialweight - finalweight \times 100}{weight - taken} \right)$$
(AOC, 1997)

4. Determination of percentage insoluble matter:

Calculation:

Percentage of insoluble matter = $\dfrac{W_1 - W_2}{P}$ X 100

$W_1 =$ Initial weight

$W_2 =$ Final weight

5. PH value determination:

The pH of ocher soap was determined using pH meter. Two grams (2g) of finished product of each soap was dissolved in 10 mls of distilled water and stirred till the sample dissolved. The probe tips were dipped inside the solution and the readings were recorded (AOC, 1997).

ORGANISM/COLONY TESTING

An already prepared, characterized and well stored isolate of *Dermatophylus congolensis* was employed as a testing organism in this study. This organism was selected because of;

> (a) its prevalence in Nigeria i.e readily available and as a causative agent of a wide range of skin infections to humans as well as animals, and

> (b) its clinical importance, needing an urgent attention most especially in children.

Antibiogram

This is the use of ochre prepared soap-extract to test against microorganism in the research laboratory. The pure isolate of the organism (*D. congolensis*) was obtained from a sister research microbiology laboratory where it was stored for research purposes at Federal University of Technology, Minna, Nigeria.

Media Preparation

1. Nutrient agar

Seven (7g) grams of the commercially prepared nutrient agar powder was weighed and dissolved in $250cm^3$ of distilled water. It was autoclaved at standard condition (121^0C for 15 minutes). The medium was cooled to about 50^0C, after the solidification, the plates were placed in the oven to dry off any available water droplets.

Plate Inoculation Method

The method of inoculation used in this study was the cork plate method. It involved the use of sterile loop to smear a growth from plate of the test organism on the surface of the medium.

The loop was sterilized with Bunsen flame and when cooled it was used to pick a loopful streaked over the plate to cover the plate surface. Holes were bored on the solidified agar using a sterile cork-borer of 4.0 mm in diameter, $0.2cm^3$ (200µl) of each soap-extract was pipetted and allowed to diffuse and incubated in an anaerobic jar for 24-48 hours. After incubation, the diameters of zone of inhibition were measured to the nearest millimeter.

RESULT AND DISCUSSSION

Result

The tables below show various observations and readings as well as qualitative analyses obtained in the course of the soap production.

Table 1: Physical Properties of the Soap Samples

Soap Sample (w.r.t)	Colour	Foaming	Texture
Palm kernel oil	Milky	Foam readily	Soft
Palm oil	Dark yellow	Foam but after shaking	Brittle
Groundnut oil	Milky	Foam easily	Brittle
Shea butter oil	Light milky	Foam readily	Brittle

w.r.t: with respect to oil

Table 2: Chemical properties of the soap samples

Soap (w.r.t)	Acid Value	Free Alkalinity	% Moisture	% Insoluble Matter	PH
Palm kernel	2.04	7.59	2.4	2.2	9.00
Palm oil	2.12	7.69	2.4	5.8	9.30
Groundnut oil	2.88	8.62	5.4	4.4	9.10

Sheabutter oil	2.76	8.04	2.8	6.6	9.40

w.r.t: with respect to.

Table 3: Antimicrobial Properties of Ochre Soap

Ochre Soap with respect to inhibitory zones: Zone diameters (mm) of inhibition of different ochre soaps based on oil types.

Testing organism	Palm oil	Palm kernel oil	Groundnut oil	Shear butter oil
Dermatophylus Congolensis	18mm	N/A	N/A	11mm

Note: N/A- No activity

Table 4: Minimum Inhibitory Concentrations (MIC) of Ochre Soap with Respect to Palm Oil

Testing organism	T1	T2	T3	T4	T5
Dermatophylus Congolensis	25mg/ml	6.25mg/ml	N/A	N/A	N/A

Table 5 Minimum Fungal Concentrations (MFC) of Ochre Soap with Respect to Palm Oil

Testing organism	T1	T2	T3	T4	T5
Dermatophylus Congolensis	25mg/ml	6.25mg/ml	N/A	N/A	N/A

Table 6: Minimum Inhibitory Concentrations (MIC) of Ochre Soap with Respect to Shear butter Oil

Testing organism	T1	T2	T3	T4	T5
Dermatophylus Congolensis	15mg/ml	3.25mg/ml	N/A	N/A	N/A

Table 7: Minimum Fungal Concentrations (MFC) of Ochre Soap with Respect to Shear butter Oil

Testing organism	T1	T2	T3	T4	T5
Dermatophylus Congolensis	15mg/ml	3.25mg/ml	N/A	N/A	N/A

DISCUSSION

Table 1 reveals variations in the colour of the soaps based on oil type. The colour observed in palm kernel oil was milky, palm oil was dark yellow, groundnut oil was milky and Shea butter oil was light milky in colour. This shows that the oil source has effect on the colour of soaps (Phillips, 2008).

More so, it was observed that the types of oil have resultant effect on the foaming characteristics of the various soaps produced (Dunn, 2010) During the analysis, it was observed that palm kernel oil soaps foamed readily, palm oil soap foamed after shaking, and ground nut oil soap foamed easily while Shea butter oil foamed readily. (Jones,2010).

Considering the texture of the soaps, it was observed that the soaps produced from palm kernel oil was soft, while soap products made of palm oil, ground nut oil and Shea butter oil were brittle; this implies that palm kernel oil was softest oil among them. (Jones, 2010).

Table 2 shows the different acid values obtained from different soap products as regards oil nature. Their acid values were 2.04, 2.12, 2.88 and 2.76 for palm kernel oil, palm oil, groundnut oil and Shea butter oil respectively. This agreed with study of (Kuntom *et al.*, 1996) in a soap manufacturing study done in Malaysia using blends of distilled fatty acid of palm oil and palm kernel oil with resultant effect of increased acid values. Again, percentages (%) of the free alkalinity of these soap products were: 7.59 for palm kernel oil, 7.69 for palm oil, 8.62 for groundnut oil and 8.04 for Shea butter oil respectively. The estimated free alkali of soap samples is within the range of the Egyptian standard (Ministry of Industry Egypt, 1992) and thus, soap made of ground nut oil showed highest percentage of free alkalinity, Shea butter oil indicated higher percentage of free alkalinity, whereas free alkalinity of both palm kernel and palm oils were lower in alkalinity and at close range (Taiwo *et al.*,2008).

The percentage moisture (%) of soap sample products for palm kernel oil, palm oil, groundnut oil and Shea butter oil were 2.4, 2.4, 5.4 and 2.8 respectively. This implies that percentage of

11

moisture content was highest in groundnut oil while both palm oil and palm kernel oil were lowest in value. The moisture content of the soap products found in the current study was lower than those obtained by (Aine *et al.*, 1996 and Mak- Mensah and Firempong *al.*, 2011). However, these results could be due to the difference in the soap preparing methods.

The percentage of insoluble matter in soap produced from these oil products namely palm kernel oil, palm oil, groundnut oil and Shea butter oil accordingly were 2.23, 5.8, 4.4 and 6.9. This present study demonstrated that palm kernel comprised highest soluble matter (%) while Shea butter oil contained highest insoluble matter (%). These values corroborate with the ranges agreed by (Eke *et al.*, 2004).

Furthermore, from the physico-chemical properties of our laboratory produced soap products, the pH values of the sample products made of palm kernel oil, palm oil, ground nut oil and Shea butter oil were 9.00, 9.30, 9.10 and 9.40 respectively. These values are in consistent with the normal pH range value for soap products of 8-10.5 (www.Soap-body-and-spa.com) and pH range of 9-11 by a separate study (Oyedele, 2002).

Finally, the minimum inhibitory concentration (MIC) of microbial testing for ochre soap products with respect to various sources of oil indicated different anti-microbial effects against *Dermatophylus congolensis*. Meanwhile, in the present study, it was demonstrated that ochre soap products with respect to palm oil and shear butter oil agents showed high MIC against the testing organism. Further analysis of these ochre soap products made of palm oil revealed the results obtained in terms of minimum inhibitory concentration (MIC) and minimum fungicidal concentrations (MFC) as shown in Tables 4 and 5.

Conclusion

In this present study, anti-fungal activity was observed from laboratory-produced ochre soap products with respect to palm oil additive by good microbial sensitivity testing suggested of possibility to offer treatment for topical fungal disease, precisely *Dermatophylus congolensis,* as against other domesticated oils. This was made possible by tailoring analytical process of the ochre soap to the conventional standards with regards to pH, texture, acid value, % free alkalinity, % moisture and % insoluble matter of these various samples respectively. Hence, if some of these ideal conditions are met, high grade of medicated ochre soap can be produced on industrial scale. Meanwhile, pH change is a possibility for differential sensitivity for oil-based ocher soap, for this factor, ocher soaps made of other oils may be a reason for failing the medicinal property attractable to palm oil product. This is because they all contained ochre substance as active ingredient in equal concentration.

References

Abu- Samura, M. T(1978). Morphological, Cultural and Biochemical Characteristic of *Dermatophilus Congolesis*. Zentralbl. Veterinarmed. B. 25: 668-688

Albrechit, R. (1974). Dermatophilus Congolesis Chronic nodular disease in man . Pediatrics 53: 907- 913

Ambrose, N. C (1996). The Pathogenesis of Dermatophilosis. Tropical Animal Health Production. 28:295-375

Burd, E.M., Juzych, L.A., Rudrik, J.T and Habib, F. (2007). Rustular Dermatitis Caused by *Dermatophilus Conglesis . Journal China Microbiology* 45: 1655- 1658

Ainie, K., Hamirun, K. and Peang- Kean, L. (1996). *Journal of American Oil Chemical Society* 73: 105- 108

Mak- mensah E.E. and Firempong, C.K (2011). Chemical characteristic of toilet soap prepared from neen seed oil. *Asian Journal of Plant Science and Research* 4: 1- 7

Oguntola, S. (2000). Solution to Mange in Rabbits, Nigeria Tribune, 22

Kuntom, A., Siew, W.L. and Tan, Y.A. (1994) *Journal of America Oil Chemical Society* 71:525- 528

Kuntom, A., Kifli, H. and Lim, P.K. (1996). Chemical and Physical Characteristic of Soap made from distilled fatty acids of palm oil and palm kernel oil. *Journal of America Oil Chemical Society* 73: 105- 108

Garzena, P and Mrina, T. (2013). The natural soap making hand book. ISBN 978-0-9874995-0-9

Isabelle, R. and Fabien, P. (2012). La Couleurs expliquee auxartistes, Editions Eyrolles, ISBN 978-2-212-13486-5.

Dunn, K.M. (2010). Scientific soap making.The chemistry of cold process, Clavicula press. ISBN 978-1-935652-09-0

AOAC (1980). Official Methods of Analysis 15th edition, Association of Official Chemist, Washington D.C

AOAC (1997). Official Methods of Analysis 15th edition, Association of Official Chemist, Washington D.C

Ministry of Industry, Egypt (1992). The annual report of the Egyptian Industrials Food Organization Cairo.

Taiwo, A. A., Oluwadare, I., Shobo, A. O., Amolegbe, S. A.(2008) *Scientific Research Essay* 3, 10 515 – 517.

Eke,U.B., Dosumu, O.O., Oladipo, E. and Agunbiade, F.O (2004). Analysis of Locally Produced Soap using shea butter oil blended with palm kernel oil. *Nigeria Journal of Science* 38: 19- 24

http://www.soap-body-and-spa.com.

Oyedele, A.O (2000). Antibacterial analysis of medicated soap. *Nigeria Journal of Natural Product and medicine* 66: 26- 29

Phillips, M (2008). Nablus olive oil soap. A Palestinian tradition lives on. Institute for middle East understanding.

Jones, G (2010). Cleanliness and Civilization. Beauty Imagined. A History of the Global Beauty Industry, Oxford University Press. ISBN 978-0-19-160961-9

Edgar, w (1995) Manufacture of soaps other detergent and glycerine pp 119-160, Ellis Horwood limited, a division of *John wiley and sons* U. S. A

George T.A (1984) Shreve's chemical process industries 5th edition pp 529-553, McGrew-Hill book company U .S .A

YOUR KNOWLEDGE HAS VALUE

- We will publish your bachelor's and
  master's thesis, essays and papers

- Your own eBook and book -
  sold worldwide in all relevant shops

- Earn money with each sale

Upload your text at www.GRIN.com
and publish for free